STEM-Engineering Process

by Patrick H. Stakem

(c) 2018

Number 12 in the STEM Series

Table of Contents

Introduction 3
Author 3
   What is STEM? 3
The Engineering Process 6
   Modular Design 7
   Optimizations 8
   Error Sources 8
   Complexity 9
   Approaches 10
Expert Opinions 11
Engineering Tools 12
   TRL 13
   Design Methodologies 15
   Root Cause Analysis 16
   FMEA 16
   Fault Tolerant Design 17
   Redundancy 18
   Safety Engineering 19
   Standards 19
Forensic Engineering 21
What's the Worst that could happen? 21
   F22 Raptor, and Y2K 21
   How not to do it – Medical devices. 22
   Mean Time to Failure for a Cathedral 23
Student Exercise 24
Bibliography 25
Resources 28
Glossary of Terms 30
If you enjoyed 32

Introduction

The book discusses the Engineering process, a road map developed over time to simplify design, implementation, and testing of systems. It is applicable across a broad range of applications, from building a cathedral, to designing ap software.

One key part of the process is that it can be enhanced and modified, to incorporated new knowledge. The key to a good process to follow is, it should define the actual problem that is to be solved.

"Anything that can go wrong, will" attributed to Murphy.

"…at the worst possible time, when you least expect it…" Anon.

"Murphy was an optimist." Anon.

Author

The author has a BSEE in Electrical Engineering from Carnegie-Mellon University, and Masters Degrees in Applied Physics and Computer Science from the Johns Hopkins University. During a career as a NASA support contractor from 1971 to 2013, he worked at all of the NASA Centers. He served as a mentor for the NASA/GSFC Summer Robotics Engineering Boot Camp at GSFC for 2 years. He teaches Embedded Systems for the Johns Hopkins University, Engineering for Professionals Program, and has done several summer Cubesat Programs at the undergraduate and graduate level.

What is STEM?

STEM (Science, Technology, Engineering, Mathematics) is the key to the United States' continued dominance in High Technology. It took a lot of expertise to implement the first cell phone. Now they are turned out like cookies in third world countries.

STEM addresses overall education policy and curriculum sources in schools, at critical grade levels.

Although the teachers are experts in their particular area, and know how to present grade-appropriate material, they may not know how to find and access the advanced resources they need, or where to get help in a particular topic area.

STEM programs are seen as critically important in the education system, world-wide. It is getting to be a complex, interconnected, global ecosystem. Advances in the subject areas of STEM will take place only by those who know how to exploit this ecosystem for knowledge.

When I was in school well before the Internet and STEM age, I had an encyclopedia, updated yearly. Today, students can access WikiPedia from their smart phones. The focus has changed from knowing facts, which are at your fingertips, on demand, to leveraging facts to innovate. This approach touches all of the academic disciplines, the Humanities, Languages, Art, besides the STEM topics. Perhaps the best skill set to have is good internet search skills. Teachers have had to transition from asking factual questions, to asking questions that derive from applications of online research, and accrued knowledge.

When I was a kid, there was no STEM. My interests in science and engineering led to research and hands-on experimentation. Luckily, I survived. I was called on, while in grade school, to demonstrate some concepts of electricity to a High School class. The first satellite was launched, and I was glued to the black & white TV. I participated in Model Rocketry at the High School Level, and went on to fly Nationally. This was made possible by an extraordinary High School Science teacher. I made quite a few friends, some of whom became Astronauts. I was given a great opportunity when I received a full scholarship to a College of my choice. I went to Carnegie Tech in Pittsburgh (now, Carnegie-Mellon University), and launched a career in Engineering and Aerospace. It is time for me to pay forward.

My thesis is, a project brings together all of the interesting pieces to provide a focal point for student work. There is a massive body of applicable free support material available. I have experience teaching engineering courses at the advanced undergraduate and graduate level, but I have no experience or credentials at the critical pre-K thru 12 levels.

I think STEM is a critical resource for understanding and implementing the future. I think the Digital I/O and Storage paradigm is a good thing to introduce into STEM. Let's do this. Future generations of STEM-mies will thank us.

Although STEM schools will have in-house expertise in Science (Physics, Chemistry, Biology), and Math (counting numbers through calculus), they are not heavily into Technology and Engineering. That's where you can ask for help – there's a lot of resources and knowledgeable individuals available out there.

The Engineering Process

The Engineering Process is a methodology or series of steps, supported by experience, that has been shown to lead to complete, correct, and functional systems. There is also a Scientific Method, which is applicable to obtaining new knowledge.

Systems Engineering, as a process, goes back to the Bell Labs in the 1940's. There was a need, given urgency by World War-II, to develop new methods applicable to new, larger problems. The co-ordination of D-Day comes to mind, as does the later Apollo missions to the moon, and the International Space Station. We might say that Systems Engineering takes a "holistic view" of the System.

A founder of modern Systems Engineering is Simon Ramo, the "R" in TRW Aerospace.

We can define a series of steps to implement a given project.

Steps of the process:

- Define the Problem Requirements – what is it we're doing?
- Do Background Research – how has this been done before?
- Develop specifications – here's exactly what we want.
- Brainstorm Solutions – kick ideas around.
- Choose the Best Solution – you may go back and revise.
- Architect the solution – What, not how.
- Design the solution – how.
- Do Development Work – build models.
- Define a test plan – How do we know it works?
- Test and Redesign – oops. It didn't work. Re-think.
- Optimize – along certain dimensions.
- Document – write it all down as a lessons-learned.

Defining the requirements is a very important first step. We need to define exactly what we are trying to do. If we don't know what we're trying to do, we won't get anywhere. The nice thing about the engineering process is that it is iterative. At any point, we can go back. Good requirements are correct, unambiguous, complete, and verifiable. Requirements come from the customer, and from required functionality. The specification and test plan is derived from the requirements. Requirements can't be contradictory. Requirements can be changed or modified, but they must be traceable back to a source. If no one said it had to be blue, then blue is not a requirement. Every specification of the systems must trace back to a requirement.

Designing the solution to a problem will probably require analysis not only of the problem, but of various solutions. We may need to build models, so we can see if our approach is good. Models can be brick and mortar, or software. Remember that certain parameters don't scale.

Optimization along certain parameters is usually required. Once we understand what it is that we want to do, and have a plan of how to do it, we may need to introduce parameters such as "it has to be built in 6 months," or "it can only draw 5 watts or power," or "it has to be waterproof to a depth of 1,000 feet."

Some optimizations don't strictly involve engineering parameters: "it can't cost more than $x," or "it has to be blue."

"Better is the enemy of good." anon.

Modular Design

There is a mind set that is useful during the design phase, and that is "how will I test this?" An approach called "design for testing" is applicable. This might be as simple as allowing access or visibility to various points or internals.

If you have to take short cuts, it is essential to document your assumptions and your violations of standard practice.

Get what ever it is that you are doing working as soon in the process as possible. It doesn't need all the parts, just the major ones. Having a model you can touch brings a reality to the project. It won't be small or fast or elegant, but it will work. Modeling of the system and its environment, as well as simulation are important in the process.

Optimizations

We can optimize for speed, size, power, time-to-complete, but not all at once. Usually, optimizing one parameter takes one or more parameters out of optimization. The key is to achieve a balance. Rank the parameters in order of importance, and work from that.

Error Sources

"Never attribute to malice what might instead be only incompetence".

Know where errors come from, historically, and you can eliminate some of them early. These are some common error sources.

- logical (design errors)
- errors due to faulty specification
- incorrect definition of the problem
- overlooking the "else" in if-then-else
- missing hidden assumptions
- incomplete scope of problem domain
- misunderstanding the requirements - are we all on the same page?
- typos, misspelling

In some systems, there is zero tolerance for errors. Think about the Apollo missions to the Moon, aircraft flight controls and air traffic control, chemical/biological/hazmat control. What about traffic lights. We assume that when ours is green, the other direction is red. Can both be green at the same time? Why not?

Complexity

One of the major sources tof errors and failures is the complexity of the project. Is there a way to measure complexity, and perhaps figure out when most errors are made? As it turns out, yes.

Complexity has two components – the number of elements, and the number of interactions between elements. Complexity applies to software as well as hardware. In a 2008 study, NASA found some answers to the complexity issues that can guide future projects.

The outcome of the study defined complexity, showed how to measure it, and how to manage it. They defined complexity as "a measure of how hard something was to understand."

They looked at the growth trends in flight software, that software that is part of the space mission. Using lines of code as a crude measure of complexity, they saw an exponential growth curve from 1968 to 2004. They also saw a similar trend in military aircraft software.

There are two types of complexity, essential, which means you are trying to solve a complex problem, and incidental, which is complexity you add in the solution. There is nothing you can do about essential complexity, but you can analyze and control incidental complexity.

They looked at complexity in several areas: requirements, implementation, testing, and operations. They targeted unintended interactions between system elements. One of their key findings was that engineers don't often realize the downstream complexity caused by their decisions. A recommendation of the study is to devote more time in analysis of requirements, to minimize the downstream complexity. Look out for unneeded features. Make sure that all requirements are testable.

One important discussion of complexity came out of the effort:

"Complexity is the label we give to the existence of many interdependent variables in a given system. The more variables and the greater their interdependence, the greater that system's complexity. Great complexity places high demands on a planner's capacities to gather information, integrate findings, and design effective actions. The links between the variables oblige us to attend to a great many features simultaneously, and that, concomitantly, makes it impossible for us to undertake only one action in a complex system. ... A system of variables is 'interrelated' if an action that affects or is meant to affect one part of the system will also affect other parts of it. Inter-relatedness guarantees that an action aimed at one variable will have side effects and long-term repercussions."

Dietrich Dörner, 1996, *The Logic of Failure*

NASA noticed a similar growth trend in Automotive software. They went on to address defects or errors at different points in the development process.

Approaches

- Be careful of "post hoc" analysis – it can be misleading
- Change only ONE THING at a time.
- Have something to observe or measure
- Apply the scientific method - postulate a hypothesis, and have a way to test it. Even negative results are results.
- Although we are looking for the simplest cause (Occam's razor, and Sherlock Holmes), multiple, simultaneous errors can occur.
- Plan for Murphy's intervention in the debugging process
- Try to break less than you fix - secondary effects. (e.g., instrumentation takes up space and time)
- There are no "non-intrusive" measurement methods. As Dr. Heisenberg said, "I'm really not sure."
- Go for the primal cause - the cause of the cause.
- Is the observed event an effect, or the effect of an effect?

Expert Opinions

"History repeats itself because no one was listening the first time." Anon.

The following are a cross-section of expert opinions, based on experience and keen judgment. You don't always need to heed the advise of experts.

"I think there is a world market for maybe five computers." - Thomas Watson, chairman of IBM, 1943.

"There is no reason anyone would want a computer in their home." - Ken Olson, president, chairman & founder of Digital Equipment Co, 1977.

"This 'telephone' has too many shortcomings to be seriously considered as a means of communication. The device is inherently of no value to us." - Western Union internal memo, 1876.

"The telephone will be used to inform people that a telegram has been sent." - Alexander Graham Bell.

"The wireless music box has no imaginable commercial value. Who would pay for a message sent to nobody in particular?" - David Sarnoff's associates in response to his urging for investment in the radio in the 1920s.

"Heavier-than-air flying machines are impossible." - Lord Kelvin, president, Royal Society, 1895.

"Professor Goddard does not know the relation between action and reaction and the need to have something better than a vacuum against which to react. He seems to lack the basic knowledge ladled out daily in high schools."- 1921 New York Times editorial about Robert Goddard's revolutionary rocket work.

"The bomb will never go off. I speak as an expert in explosives." - Admiral William Leahy, US Atomic Bomb Project.

"This fellow Charles Lindbergh will never make it. He's doomed." - Harry Guggenheim

"Airplanes are interesting toys but of no military value." - Marechal Ferdinand Foch, Professor of Strategy, Ecole Superieure de Guerre.

"Man will never reach the moon regardless of all future scientific advances." Dr. Lee De Forest, inventor of the vacuum tube.

"Everything that can be invented has been invented." - Charles H. Duell, Commissioner, U.S. Office of Patents, 1899.

"640K (memory) ought to be enough for anybody." - Bill Gates, 1981.

"The Transistor is a passing fad."- Dr. William J. Barclay, EE Department, NCSU, 1969.

Engineering Tools

"There is no bad situation that you can't make worse." attributed to Astronaut Corps.

Various system engineering tools are available to use both before and after "incidents." Ideally, we review the systems before hand with respect to failure and safety, and factor these issues into the implementation plan. Many factors marginalize this approach, including management focus, and impact on systems cost and schedule.

After the fact, we have the Root Cause analysis method, so we can determine what exactly failed, and how this particular issue can be addressed and mitigated. In many cases, this process uncovers other latent issues that also need to be addressed. These need to be documented as case studies for the use of future projects.

One of the most critical Engineering tools in systems engineering is the notebook. Write everything down, as it is happening. Jot down what certain design decisions were made. Do "lessons learned" in real time. I like the composition books, that are graph-ruled. Anything will work. You can't write too much. Analyze and understand how you make mistakes.

TRL

The Technology readiness level (TRL) is a measure of a device's maturity for use. There are different TRL definitions by different agencies (NASA, DoD, ESA, FAA, DOE, etc). TRL are based on a scale from 1 to 9 with 9 being the most mature technology. The use of TRLs enables consistent, uniform, discussions of technical maturity across different types of technology. We will discuss the NASA one here, which was the original definition from the 1980's.

Technology readiness levels in the National Aeronautics and Space Administration (NASA)

1. Basic principles observed and reported
This is the lowest "level" of technology maturation. At this level, scientific research begins to be translated into applied research and development.

2. Technology concept and/or application formulated
Once basic physical principles are observed, then at the next level of maturation, practical applications of those characteristics can be 'invented' or identified. At this level, the application is still speculative: there is not experimental proof or detailed analysis to support the conjecture.

3. Analytical and experimental critical function and/or characteristic proof of concept.

At this step in the maturation process, active research and development (R&D) is initiated. This must include both analytical studies to set the technology into an appropriate context and laboratory-based studies to physically validate that the analytical predictions are correct. These studies and experiments should constitute "proof-of-concept" validation of the applications/concepts formulated at TRL 2.

4. Component and/or breadboard validation in laboratory environment.

Following successful "proof-of-concept" work, basic technological elements must be integrated to establish that the "pieces" will work together to achieve concept-enabling levels of performance for a component and/or breadboard. This validation must be devised to support the concept that was formulated earlier, and should also be consistent with the requirements of potential system applications. The validation is "low-fidelity" compared to the eventual system: it could be composed of ad hoc discrete components in a laboratory

TRL's can be applied to hardware or software, components, boxes, subsystems, or systems. Ultimately, we want the TRL level for the entire systems to be consistent with our flight requirements. Some components may have higher levels than needed.

5. Component and/or breadboard validation in relevant environment.

At this level, the fidelity of the component and/or breadboard being tested has to increase significantly. The basic technological elements must be integrated with reasonably realistic supporting elements so that the total applications (component-level, sub-system level, or system-level) can be tested in a 'simulated' or somewhat realistic environment.

6. System/subsystem model or prototype demonstration in a relevant environment (ground or space).

A major step in the level of fidelity of the technology demonstration follows the completion of TRL 5. At TRL 6, a representative model or prototype system or system - which would go well beyond ad hoc, 'patch-cord' or discrete component level breadboarding - would be tested in a relevant environment. At this level, if the only 'relevant environment' is the environment of space, then the model/prototype must be demonstrated in space.

7. System prototype demonstration in a space environment.

TRL 7 is a significant step beyond TRL 6, requiring an actual system prototype demonstration in a space environment. The prototype should be near or at the scale of the planned operational system and the demonstration must take place in space.

The TRL assessment allows us to consider the readiness and risk of our technology elements, and of the system.

Design Methodologies

These are processes for creating a system. Ideally, processes that have worked in the past. As systems get more complex, a rigorous approach to their implementation is required. We have to do the engineering, but also meet a desired time to complete, control costs, and meet quality standards. A design flow is the series of steps in a design methodology. For large projects, there are software tools to keep track of the steps.

One approach is the successive refinement model. Here, we get something working early, and then refine it incrementally. Generally, project teams have people from different disciplines. This can speed up development and testing, but only if every one is on the same page.

Feedback is essential in every aspect of the project. Frequent team meetings help. It is the Team leader's job to keep every one focused on the big picture, and to resolve issues.

Root Cause Analysis

Root Cause Analysis (RCA) refers to an engineering process to identify and categorize the causes of events, and to identify the primal cause. It is a useful tool for determining why a disaster happened. It is used to define the what, how, and why (and sometimes, who). Its value is that it will lead to a definition of corrective measures that can be applied in the future. By definition, root causes are underlying, identifiable, and controllable.

The RCA process includes a data collection phase (forensics), a cause charting, the root cause identification, recommendations, and implementation of the solution to avoid repeating the error. In many cases, the RCA will uncover other failure causes that were overlooked.

FMEA

The failure modes and effects analysis is an engineering tool that is applied during the design and testing process of a system. In this approach, we postulate failure modes, and analyze their impact on the system performance. The possible failure modes are examined to confirm their validity. Then, the possible failures are prioritized by severity and consequences. The goal is to identify and eliminate failures in the order of decreasing severity.

The FMEA approach can actually start at the Project conceptual phase, and continue throughout the project life-cycle. It can (and should) be applied to modifications to existing projects.

The origins of the FMEA approach were during World War-II, by the U. S. Military. After the war, the approach was adopted by the aviation (aerospace) and automotive industries.

The FMEA analysis requires a cross-functional team, consisting, as applicable of hardware and software engineers, manufacturing, Quality Assurance, test engineers, reliability engineers, parts engineers, and, ideally, the customer.

The process involves identifying the scope of the project, defining the boundaries and the desired level of detail. Then, the system (or project) functions are identified. Each function is analyzed to identify how it could fail. For each of these failure cases, the consequences are noted. These range from no effects to catastrophic.

Formally, the consequences are rated on a scale of 1 to 10, with 1 being insignificant to 10, catastrophic. The root cause is then determined for each consequence, starting with the 10's. Software tools are available to support this analysis process.

Once the causes are determined, the controls are defined. Controls prevent the cause from happening, reduce the probability of happening, or detect the failure in time for correction to be applied. For each control, then, a detection probability rating is calculated (or estimated), again on a scale of 1-10. Here, 1 indicates that control is certain, and 10 indicates that the solution will not work. By definition, critical characteristics of the system have a severity of 9 or greater, and have an occurrence and detection rating of greater than 3.

A Risk Priority Number (RPN) is calculated, which is severity times occurrence time detection (ratings). This measure is used to rank failure modes in the order in which they are to be addressed. Of course, some of these rankings are not measurable, but the result of good engineering guesses.

Fault Tolerant Design

In this design approach, a system is designed to continue to operate properly in the event of one or more failures. It is sometimes referred to as graceful degradation. There is, of course, a limit to the number of faults or failures than can be handled, and the faults or failures may not be independent. Sometimes, the system will be designed to degrade, but not fail, as a result of the fault. Fault recovery in a fault-tolerate design is either roll forward, or roll back. Roll back refers to returning the system state to a previous checkpointed state. Roll forward corrects the current system state to allow continuation.

Redundancy

Redundancy refers to the technique of having multiple copies of critical components. Belt and suspenders; two independent ways to accomplish the same results. Either can fail without affecting the other. This can refer to hardware or software. This increases the reliability of the system. Redundant units can be deployed in parallel, such as extra structural members, where each single unit can handle the load. This provides what is referred to as a margin of safety.

In certain systems that are responsible for safety-critical tasks, we might triplicate the critical portion, which, reduces the probability of system failure to small, acceptable, levels. This approach is found in aircraft controls, nuclear power plant controls, and many more safety-critical systems.

Of course, if there is a common error in the three units, we have not increased our reliability. This situation is referred to as a common mode or single point error. Another problem is in the voting logic, that makes the decision that an error has been made, and switches controllers. At least one satellite launch failed because the voting logic made the wrong choice. It made a mistake thinking it had made a mistake.

Redundancy carries penalties in size, weight, power, cost, and testing complexity. Fault isolation allows the system to operate around the failed component, using backup or alternative modules. Fault containment strives to isolate the fault, and prevent propagation of the failure.

One principle of fault tolerant design is replication, with multiple copies of critical systems. Of course, this approach is susceptible to common mode errors. A more vigorous approach is Diversity, where the same task may be accomplished with different implementations. This was the approach chosen for the Shuttle's computers. Generally, a fault detection systems needs to be a separate, independent entity. It's probably of error will be smaller than the main system, because it is simpler. But, it must be closely examined for common mode faults, such as a shared power supply. Systems can be designed to be fail-safe, fail-soft, or can be "melt-before-fail." The more fault tolerant that is built into a system, the more it will cost, and the more difficult it will be to test. It is important not to increase the complexity to the point where the system is not testable, and is "designed to fail."

Safety Engineering

Safety Engineering is a systems engineering approach to providing safety in systems, starting from the very beginning, and continuing through the testing phase. Safety Engineering has the goal of preventing hazards and failures.

Standards

There are many Standards applicable to systems. These range from general computer standards to embedded-specific standards. Why should we be interested in standards? Standards represent an established approach, based on best practices. Standards are not created to stifle creativity or direct an implementation approach, but rather to give the benefit of previous experience. Adherence to standards implies that different parts will work together. Standards are often developed by a single company, and then adopted by the relevant industry. Other Standards are imposed by large customer organizations such as the Department of Defense, or the automobile industry. Many standards organizations exist to develop, review, and maintain standards.

Standards exist in many areas, including hardware, software, interfaces, protocols, testing, system safety, security, and certification. Standards can be open or closed (proprietary). Sometimes, the customer will require an adherence to specified standards.

Hardware standards include the form factor and packaging of chips, the electrical interface, the bus interface, the power interface, and others. The JTAG standard specifies an interface for debugging.
In software, an API (applications program interface) specifies the interface between a user program, and the operating system. To run properly, the program must adhere to the API.

There are numerous Quality standards, such as those from ISO, and Carnegie-Mellon's CMM (Capability Maturity Model). CMM defines five levels of organizational maturity in a company or institution, and is independently audited. Language standards also exist, such as those for the ANSI c and Java languages. Networking standards include TCP/IP for Ethernet, the CAN bus, and USB.

The ISO-9000 standard was developed by the International Standards Organization, and applies to a broad range of industries. It concentrates on process. It's validation is based on extensive documentation of organization's process in a particular area, such as software development, system build, system integration, and test and certification.

It is always good to review what standards are available and could be applied to a particular system, as it ensures the application of best practices from experience, and interoperability with other systems.

Forensic Engineering

Forensic Engineering is the discipline that looks into the determination of causes, after the disaster has occurred. This data, hopefully, will be including in engineering best practices. It also provides data for court cases related to the incident. Early forensic investigations were made of 19th century bridge collapses (We're still not getting that right).

In one case in 1847, a train fell through a bridge that had been designed by famed engineer Robert Stephenson. It went to inquest, and was determined to be Stephenson's fault.

There is a systematic way to identify risks, and to reduce or mitigate them. This is referred to as Disaster Risk Reduction. It is based on research, and past case studies. The UN Office for Disaster Risk Reductions says DDR is, "The conceptual framework of elements considered with the possibilities to minimize vulnerabilities and disaster risks throughout a society, to avoid (prevention) or to limit (mitigation and preparedness) the adverse impacts of hazards, within the broad context of sustainable development." The UN addresses large humanitarian disaster, but their defined approach reaches across all instances and disciplines. (https://www.unisdr.org/)

What's the Worst that could happen?

The author has done two books on this topic, and has enough material for several more. Let's just look a a few selected cases.

F22 Raptor, and Y2K

Built by Lockheed Martin, the F-22 Raptor is an advanced stealth U. S. fighter jet. It entered service in 2005, and now has mostly been replaced by the F-35.

In February 2007, on the aircraft's first overseas deployment from Hawaii to Japan, a 15 hour flight, six F-22s of the 27th Fighter Squadron flying from Hawaii, experienced multiple software-related system failures while crossing the International Date Line. They lost navigation, communications, and some other onboard systems. Luckily, they were refueling at the time, and the refueling planes stayed with them, escorting them back to Hawaii. Ooops. The aircraft landed ok, and the software error error was fixed in a couple of days

How not to do it – Medical devices.

(this is courtesy of a student in my Embedded Systems class, at JHU.)

A leading manufacturer of ventilation products for the Hospital and pre-hospital markets with and annual revenue of 9.6 Billion Euros, invests 8% of that revenue into the R&D efforts. In 2017 the company was forced to issue a recall on some of its products due to a software problem. The recall affected all of the gold standard hospital dedicated noninvasive ventilator with version 2.20 software release. The software defect could erroneously report the blower motor has stalled and cause the unit to shut down leading to hypoxia due to lack of oxygen. Despite the company's effort to make this product safe, the FDA declared this failure to be a Class I recall (the highest-risk label on a recall) that affects more then 19,000 units.

The fault is falsely triggered but the software does behave appropriately to the failure it wrongly detected; once detected the device will displaying an "E100 error code", and shut down while alarming in a high priority alarm mode. The company is a leader in the ventilator space and this recall was determined to "produce a situation in which there is a reasonable probability that the use of or exposure to the product will cause serious adverse health consequences or death." Coincidentally, there was a recall a year earlier for potential motor failures that affected 116 units and was related to an issue in the blowers ventilator assembly.

Evidently, pacemaker software is not known for its reliability but is known for its vulnerability to hacking. A recent study defined hundreds of known vulnerabilities in the code. A very small percentage of manufacturers address security, making their devices wide open to cyber attack. There is also supposed to be a login code and password, for when the device needs an upgrade. Even this was ignored. Less than half of the manufacturers were found to be following advise on device security.

Mean Time to Failure for a Cathedral

The Gothic style Beauvais Cathedral in 1284 France suffered a major collapse of its vaulting in the choir. This started a new thinking about cathedrals, and introduced a atmosphere of fear and conservatism among architects. The collapse may have been caused by wind gusts. There was little understanding of buildings and structural mechanics since the Greek and Romans. The Cathedrals were meant to be vast inside, with your eyes drawn up to the heavens. Unfortunately, doing this in stone had reached its limits, even with exterior buttresses. Stone is good in compression, but a bad building material in tension – it can literally be pulled apart. It was not until the advent of iron structures that tension members could be used. In some cases, a large chain was wrapped around the top of the building to take the tension forces. Today, the repairs on Beauvais are not yet complete, and laser scanners and 3-D models are being used to analyze the structure.

The National Cathedral and the Washington Monument in Washington, D. C. both suffered damage due to a rare earthquake in the area in 2011.

We should also mention the Leaning Tower or Pisa. Some one underestimated the weight, and the softness of the ground. It actually started to lean during construction in 1173. Probably built by the low bidder. Major restoration work took place in 1990, reducing the angle of the lean from 5.5 to just under 4 degrees. It should be good to go for another 300 years.

Student Exercise

It is useful to apply some of these techniques in the STEM class, starting with "paper projects." Nothing has to be built at first, but the process should be applied to a proposed project. It is useful to break the class into Design Teams of 4-5. The teacher postulates a project, and defines a set of requirements. The Design Teams respond to these requirements with a specification, and a test plan. Included should be an estimate of complexity, and an estimated time to complete.

Each Team does a Presentation before the rest of the class (members of the Preliminary Design Review Board). The teacher participates in this only as an observer. Let this be a peer review. The teacher can then grade both the specification and the test plan.

It is essential that the project be "real-world" and include as much engineering and math as possible. The students should be challenged to do the background research. This will be a some what like a Science Fair Project.

Along the way, the Review Board will see items that missed in developing their own materials – this is ideal, learning from some one else's mistakes or omissions. The Projects will give the students firsthand experience in applying science, math, engineering, and proper processes to a problem. It will enhance a collaborative working environment. The students will also experience a presentation to their peers.

Bibliography

Anderson, Ross J. *Security Engineering: A Guide to Building Dependable Distributed Systems*, 2008, ISBN-978-0470068526.

Blockley, David; Godfrey, Patrick *Doing it Differently: Systems for rethinking infrastructure*, 2017, ICE Publishing, ISBN-0727760823.

Buede, Dennis M.; Miller, William D. *The Engineering Design of Systems: Models and Methods*, 3rd ed, John Wiley and Sons, 2016, ISBN-111902790X.

Cheng, P.G. *100 Questions for Technical Review,* Aerospace Report No. TOR-2005(8617)-4204. Space and Missile Systems Center. September 30, 2005.

de Weck, Olivier L.; Roos, Daniel, *Engineering Systems: Meeting Human Needs in a Complex Technological World*, 2016, ISBN-0262529947.

Dornet, Dietrich *The Logic Of Failure: Recognizing And Avoiding Error In Complex Situations*, 1997, ISBN-0201479486.

Durant, Will and Ariel, *The Lessons of History*, 1st ed, 1968, Simon & Schuster, ISBN- 143914995X.

Fawcett, Bill *Trust Me, I Know What I'm Doing: 100 More Mistakes That Lost Elections, Ended Empires, and Made the World What It Is Today,* Berkley, 2013, ISBN-0425257363.

Gill, Paul S. ; Garcia, Danny *Engineering Lessons Learned and Systems Engineering Applications,* NASA, avail: https://www.researchgate.net/publication/242185667_Engineering_Lessons_Learned_and_Systems_Engineering_Applications

Griffin, Paul M., Nembhard, Harriet B. *Healthcare Systems Engineering*, 2016, ISBN-1118971086.

Hermann, Debra S. *Software Safety and Reliability: Techniques, Approaches, and Standards of Key Industrial Sectors*, Wiley-IEEE Computer Society Press; 1st edition, February 10, 2000, ISBN0769502997.

Krantz, Gene (2001). *Failure Is Not an Option: Mission Control from Mercury to Apollo 13 and Beyond,* New York: Simon & Shuster. ISBN 978-0-7432-0079-0.

Levenson, Nancy G., Moses, Joel *Engineering a Safer World: Systems Thinking Applied to Safety*, 2012, ISBN-

Maier, Mark W. *The Art of Systems Architecting*, Third Edition, 2009, ISBN-

Meadows, Donella H., Wright, Diana (Ed) *Thinking in Systems: A Primer,* Chelsea Green Publishing, ISBN-1603580557.

NASA, Systems Engineering Handbook: NASA/SP-2016-6105 Rev2, 2017, ASIN-B076MH5VGT.

Petroski, Henry *To Forgive Design: Understanding Failure*, Belknap Press of Harvard University Press, 2012, ISBN-10: 0674065840.

Petroski, Henry *Success through Failure: The Paradox of Design,* Vintage, 1992, ISBN-0679734163.

Ryan, R. S. "A History of Aerospace Problems, Their Solutions, Their Lessons," NASA Technical Paper 3653, 1996, avail:
https://ntrs.nasa.gov/archive/nasa/casi.ntrs.nasa.gov/19970001339.pdf

Spark, Nick T. *A History of Murphy's Law,"* Periscope Film, 2006, ISBN 0-9786388-9-1.

Stakem, Patrick H. *What's the Worst that could Happen? Bad Assumptions, Ignorance, Failures, and Screw-ups in Engineering Projects,* 2016, ISBN-1520207166.

Stakem, Patrick H. *What's the Worst that could Happen? More Bad Assumptions, Ignorance, Failures, and Screw-ups in Engineering Projects, 2018 ,* Volume-II, ISBN-1981005579.

Stevens, Richard, Brook, Peter System Engineering: Coping with Complexity, 1998, Prentice Hall, ISBN-0130950858

Sweet, Justin; Schneier, Marc M. *Legal Aspects of Architecture, Engineering and the Construction Process*, 9th Edition, 2012, ISBN-1111578710.

U. S. Army, *Systems Engineering Fundamentals*, 2013, ISBN 1484120833.

Wasson, Charles *System Engineering Analysis, Design, and Development: Concepts, Principles, and Practices*, 2015, Wiley, ISBN-1118442261.

Resources

- The Engineering Design Process: The 4 Key Steps to STEM Teaching and Learning. Avail: https://www.advancementcourses.com/blog/the-engineering-design-process-the-4-key-steps-to-stem-teaching-and-learning/

- *Systems Engineering: Challenging Complexity,* the Open University, avail: http://www.open.edu/openlearn/science-maths-technology/computing-and-ict/systems-computer/systems-engineering-challenging-complexity/content-section-0

- NASA Flight Software Complexity, avail: https://www.nasa.gov/pdf/418878main_FSWC_Final_Report.pdf

- https://www.nasa.gov/audience/foreducators/best/edp.html

- https://www.khanacademy.org/partner-content/49ers-steam/49ers-gridiron-eng/49ers-innovations-equipment/v/engineering-design-process

- https://theworks.org/educators-and-groups/elementary-engineering-resources/engineering-design-process/

- https://www.nationalgeographic.org/media/engineering-process/

- Merlin, Peter W., Bendrick, Gregg A. NASA History Program Office *Breaking the Mishap Chain: Human Factors Lessons Learned from Aerospace Accidents and Incidents in Research, Flight Test, and Development*

(NASA Aeronautics Book), 2013.

- Levenson, Nancy sunnyday.mit.edu/accidents/

- http://www.films.com/id/9417/When_Engineering_Fails.htm (51 minute video)

- NASA Systems Engineering Handbook, NASA/SP-2007-6105 Rev1, December 2007, avail:

https://ntrs.nasa.gov/archive/nasa/casi.ntrs.nasa.gov/20080008301.pdf

- The Mitre System Engineering Guide, avail: https://www.mitre.org/publications/technical-papers/the-mitre-systems-engineering-guide

- http://standards.nasa.gov

- http://www.icseng.com/

- System Engineering Body of Knowledgeable https://incoseonline.org.uk/Normal_Files/Home/Default.aspx?CatID=Home

- http://www.systemsengineeringtool.com/

- Office of the Deputy Assistant Secretary of Defense, Systems Engineering, avail: http://www.systemsengineeringtool.com/

- US DoD MIL-STD-499 System Engineering Management

- NASA Lessons Learned Information System, avail: http://llis.nasa.gov

Glossary of Terms

CPSC – U. S. Consumer Product Safety Commission.
DFD – Data Flow Diagram
Error segmentation – keeping the effects of an error from cascading.
Fail-safe – a failure does not cause a fault.
Fault Tolerant – a property of a system where 1 or more faults won't cause it to fail.
Fault Tree – a graphical representation of faults and causes.
FFBD – functional flow block diagram.
FMEA – Failure Modes and Effects Analysis
HFE – Human Factors Engineering
Hubris – from the Greek, extreme pride or self-confidence; sometimes found in engineering design.
INCOSE – International Council on Systems Engineering.
IV&V – independent verification and validation – having some one else look over the system.
JTAG – Joint Test Action Group.
Life Cycle – from the very beginning to the very end.
LLIS – (NASA) lessons learned information system
MTBF – Mean time between failures
MTTF – Mean Time to failure
MTTR – Mean time to Repair.
Murphy's Law – anything that can go wrong, will.
NASA – National Aeronautics and Space Administration.
NCOSE – National Council on Systems Engineering.
NCSU – North Carolina State University
Plan B – what to resort to when Plan A fails.
Post-mortem – investigation after the fact.
PRA – Probabilistic Risk Assessment.
Redundant – providing multiple units, either completely identical; or functionally identical.
RCA – Root Cause Analysis
Reset – return to a known, initial condition.
Root cause - the first cause, the starting point of events.

SEBoK – System Engineering Body of Knowledge.
SE&I – Systems Engineering & Integration
SI – System Internationale (metric)
Triplicate – provide three solutions.
UML – Universal Modelling Language
Voting logic – choose the majority, based on the assumption that 2 failures are less probable than one.

If you enjoyed

Stakem, Patrick H. *Enviro-Bots for STEM, Using Robotics in the pre-K to 12 STEM Curricula, A Resource Guide for Educators,* 2017, PRRB Publishing, ISBN-9781549656613.

Stakem, Patrick H. *STEM-Sat, Using Cubesats in the pre-K to 12 STEM Curricula, A Resource Guide for Educators,* 2017, PRRB Publishing, ISBN-1549656376.

Stakem, Patrick H. *Introducing Astronomy in the Pre-K to-12 STEMCurricula, A Resource Guide for Educators,* 2017, PRRB Publishing,

Stakem, Patrick H. *Introducing Computer Math for Pre-K to 12 STEM, A Resource Guide for Educators,* 2017, PRRB Publishing,

Stakem, Patrick H. *Introducing Weather in the pre-K to 12 STEM Curricula, A Resource Guide for Educators,* 2017, PRRB Publishing,

Stakem, Patrick H. *Cubesat Engineeering*, PRRB Publishing, 2017, ASIN- B01N4VC99B.

Stakem, Patrick H. *Cubesat Operations*, PRRB Publishing, 2017, ASIN- B01N18JXX6.

Stakem, Patrick H. *Interplanetary Cubesats*, PRRB Publishing, 2017, ASIN- B06XG4DMVW.

Stakem, Patrick H. *Cubesat Constellations, Clusters, and Swarms*, PRRB Publishing, 2017, ASIN- B06X3SLRFT.

Stakem, Patrick H. *Visiting the NASA Centers and Historic Rockets & Spacecraft,* 2017, PRRB Publishing, ASIN- B0757ZVB2G.

www.ingramcontent.com/pod-product-compliance
Lightning Source LLC
Chambersburg PA
CBHW031517210526
45464CB00007B/2950